Common Sense Guide to Environmental Management

Subash Ludhra

Routledge
Taylor & Francis Group
LONDON AND NEW YORK

First published 2014
by Routledge
2 Park Square, Milton Park, Abingdon, Oxon, OX14 4RN

and by Routledge
711 Third Avenue, New York, NY 10017

Routledge is an imprint of the Taylor & Francis Group, an informa business

British Library Cataloguing in Publication Data
A catalogue record for this book is available from the British Library

Library of Congress Cataloging-in-Publication Data
Ludhra, Subash, author.
Common sense guide to environmental management / Subash Ludhra. — First edition.
p. ; cm.
Includes bibliographical references and index.
I. Title. [DNLM: 1. Environmental Health.
2. Environmental Policy. 3. Safety Management. WA 30.5]
RA418
362.1—dc23 2013039463

ISBN13: 978-0-415-83541-1 (pbk)
ISBN13: 978-1-315-85877-7 (ebk)

Typeset in Sabon
by Keystroke, Station Road, Codsall, Wolverhampton
Printed and bound in Great Britain by Ashford Colour Press Ltd

MIX
Paper from responsible sources
FSC® C011748
FSC
www.fsc.org

Contents

Foreword

As someone who has lived in the countryside and worked on a farm all my life, I have a close affinity to the environment and fully understand the vital importance of protecting it for future generations. The large farming group of which I am Chief Executive Officer has a great part to play in maintaining healthy soils (living worlds in themselves), woodlands, streams and waterways. These all support a balanced wildlife and vibrant biodiversity, without which we do not have a future.

Protecting our environment, both now and into the future, is of critical importance for long-term sustainability. We must look after our valuable assets and, in any case, preserving resources makes very good sense. Well-managed organisations protect, preserve and sustain, and the disciplines of these good management processes serve us well in all aspects of our endeavours.

When environmental accidents happen, the negative impact on the image of the business is almost too difficult to calculate. Similarly, it is hard to gauge the effect such events have on a company's ability to function in the short, medium and long term.

All of this, of course, is in addition to consideration of damage to habitats, the environment and the people living within it.

Good management of the environment equates to good management in general, and who would not want to achieve that? Education is vital to ensure that the good practices that need to be implemented are undertaken on an ongoing basis.

This *Common Sense Guide to Environmental Management* is for those with little or no environmental expertise; it is written to support training and provides an easy reference for learning about environmental management. The guide assumes no prior knowledge and is easy to use.

It therefore provides a valuable part of this series of common sense guides to increase understanding and it follows that any improved learning gives us all an even greater chance to take the actions needed today and tomorrow to protect the future.

I was delighted to be asked by the author to write this foreword, and I wish you all well as you strive each and every day to protect your environment.

John Shropshire – CEO
The Shropshire Group

Welcome

It is often said that we all know what is right and what is wrong, what we should and should not do. Surely it's just "common sense". Unfortunately common sense is not as common as we would like to believe. The definition of common sense is "the ability to behave in a sensible way and make practical decisions". Most individuals and employers will at some time do things that are not sensible or practical and some will do this regularly.

This guide has been developed to help improve your knowledge of environmental issues and how to manage them within the workplace in a light-hearted way. It is designed to further heighten the common-sense element of your existing knowledge.

Although the guide refers to UK legislation/best practice the principles are applicable internationally. However, you should answer all questions within the guide, based on the UK legislation.

By taking the time to improve your knowledge and learn more about environmental management in the workplace you can:

- **avoid directly polluting the environment;**
- **help prevent incidents where others may pollute the environment;**
- **make your workplace more energy efficient;**
- **apply the same practices within your home.**

Every year, thousands of environmental incidents take place. Fortunately most are relatively minor but occasionally some have international impact and remain in our minds for a long time. Sadly some types of pollution are occurring constantly and have become almost accepted as necessary evils of our lifestyles (e.g. carbon dioxide emissions) although greater efforts are being made to tackle them.

The damage already done through hundreds of thousands of years of environmental pollution (natural and man caused) can probably never be reversed. However, we do have an opportunity to slow down the process of pollution and in some cases cease it. However, it will only happen if everybody plays their part. All of our actions (however small or insignificant) can collectively make a huge impact on reducing global environmental pollution, which will in turn have a positive impact on the quality of our lives.

DID YOU KNOW?

- Pollution is one of the biggest global killers, affecting over 100 million people.
- Over 1 billion people worldwide lack access to safe drinking water. Five thousand people die each day due to dirty drinking water.
- Over 1 million seabirds and 100,000 sea mammals are killed by pollution every year.
- More than 3 million children under age five die annually from environmental factors.

By working through this accessible guide you will learn more about environmental pollutants and how they can be better managed within the workplace and home, and as a result you will be better placed to recognise the hazards and dangers in any workplace or even your own home 24 hours a day.

How this guide works

AIMS

The aim of this guide is to provide you with a basic understanding of:

- the need to protect the environment;
- environmental law and enforcement;
- environmental legislation;
- environmental incidents and what causes them;
- environmental hazards within the workplace; and
- proactive measures that can be taken to help reduce pollution and waste or the impact of it and minimise the use of energy.

OBJECTIVES

By the time you finish this guide, you will be able to:

- explain why it is important to protect the environment;
- define the terms environmental aspect and impact;
- describe the greenhouse effect and its implications;
- describe the main pollutants of air, their effects and controls available;
- describe the main pollutants of water, their effects and controls available;
- describe the main pollutants of land, their effects and controls available;
- explain why waste must be managed and controlled;
- have a greater understanding of environmental management and its importance in your everyday work and home life;
- list examples of renewable and non-renewable energy sources;

- explain why we need to conserve energy;
- give examples of different environmental management systems;
- explain the need for planning for emergencies.

How to complete the guide

Before you start to complete this guide please read the notes below in order to ensure that you get the most out of your training.

WHAT DO YOU NEED TO COMPLETE THE GUIDE?

You will need:

- a quiet cosy environment that allows you to relax and make notes;
- a pen and paper to make notes;
- a desire to learn and improve your knowledge of environmental management.

GUIDANCE ON LEARNING

This guide has been produced to help you learn more about environmental management within your workplace and to reduce the likelihood of you or your employer causing environmental pollution, whilst getting you to think more about conservation of natural resources. The guide allows you to complete your studies at your own pace with the support of your manager or supervisor. However we recommend that you complete the guide within four weeks.

Throughout the guide there are simple questions designed to help you test your subject knowledge and learn.

If you cannot answer the questions please read the relevant topic again to refresh your memory. If you are still in doubt please speak to your line manager who will be able to assist you.

When you have completed the guide and the exercises, you may wish to complete the examination (20-question multiple-choice exam paper on pages 107–13). On achieving the required pass mark (the pass mark is 75%) you will receive a certificate to confirm that you have completed the guide and passed the associated examination.

HOW TO USE THE GUIDE

The guide is divided into modules. We recommend that you complete one module at a time in full, starting with module one, progressing sequentially through to the last module. You do not have to complete the guide in one sitting. A lot of information is provided and you may learn more effectively by tackling the modules in bite-sized chunks.

The modules are ordered to take you through a specific learning pattern to help you learn. Each module contains questions to make you think about your own job and workplace. There are also questions at the end of each module to test your knowledge and understanding of it. The answers can be found on pages 95–101 of the guide. There may be times when you feel you need help and support in completing the guide. Should this be the case please speak to your line manager.

WHEN DO I GET MY CERTIFICATE?

Once you have completed the guide and the examination paper, the paper will be marked and on achieving the required pass mark a certificate pdf will be emailed to you as soon as possible.

Remember the certificate only confirms that you have completed the guide and passed the associated exam. The real benefit to you will come from your improved knowledge and ability to identify potential pollution situations and improve the environment around you. **Good Luck!**

Winning trophy

MODULE ONE

The environment and environmental law

BUNCEFIELD: BRITAIN'S £1 BILLION DISASTER . . .

Welcome to module one. In this module you will learn about the importance of protecting the environment and the Environmental Protection Act.

WHY PROTECT THE ENVIRONMENT?

1. To comply with statutory legislation

You and your employer are required to comply with the requirements of the Acts and Regulations relevant to your employer's business (this will be covered in more detail later). Examples of environmental Acts and Regulations include:

- Environmental Protection Act;
- The Waste Regulations;
- The Producer Responsibility Obligations (Packaging Waste) Regulations;
- The Waste Electrical and Electronic Equipment Regulations;
- The Air Quality Standards Regulations;
- The Environmental Protection (Controls on Ozone-Depleting Substances) Regulations;
- The Clean Air Act.

This is only a small selection of the Acts and Regulations that companies have to comply with.

2. To comply with internal company policies

Responsible employers have internal environmental policies and procedures. These policies and procedures set out their own internal standards, which in some cases may exceed the requirements of the relevant regulations. You, as an employee, agree to comply with these as part of your terms and conditions of employment.

3. Moral obligations

To reduce the number of polluting incidents and to safeguard the environment and you the employee.

Employees do not generally set out to pollute the environment or waste energy. Therefore by increasing your own awareness of environmental management you can help reduce the likelihood of pollution incidents occurring and safeguard the well being of the environment.

4. It's cost effective

All pollution incidents cost money, for example clean up costs, damaged reputation, compensation payments and fines. However, the real cost is the damage to the environment in the short and longer term, which may be irreparable.

There can also be an impact on the aesthetics of the affected area and the longer-term impact on future generations.

5. There is a business need to do so

To maintain the employer's good reputation, more and more companies require their suppliers to have robust environmental management systems in place and to be able to demonstrate the promotion of pollution prevention and energy conservation as part of a wider corporate social responsibility agenda.

> "A POOR ENVIRONMENTAL REPUTATION COULD COST YOUR EMPLOYER'S BUSINESS DEARLY."

We must all work together

Q

What policies and procedures does your employer have to deal with the environment?

What is the environment?

The environment can mean our immediate surroundings at work or at home; it can even mean our town, city, county, country right up to our planet. It consists of three media, these being air, land and water.

We need to understand that all parts of the environment are inter-connected and an adverse impact on one part can affect others. Our actions, however small, can have a huge short- or long-term impact globally, for example the radiation from Chernobyl was detected in Wales (UK).

What is pollution?

The release from any process or substance into the environment (any medium) which is capable of causing harm to man or any living organism supported by the environment can be defined as pollution.

> "THE INTRODUCTION INTO THE ENVIRONMENT OF SUBSTANCES OR ENERGY LIABLE TO CAUSE HAZARDS TO HUMAN HEALTH, HARM TO LIVING RESOURCES AND ECOLOGICAL SYSTEMS, DAMAGE TO STRUCTURES OR AMENITY OR INTERFERENCE WITH THE LEGITIMATE USE OF THE ENVIRONMENT."

Pollution can include noise, heat, vibrations or any other release of energy.

Sustainable development

Many responsible organisations are taking a lead in developing products or providing services which are deemed to be sustainable. This means being able to "meet the needs of the current generation without compromising the ability of future generations to meet their own needs". For example, this may be through the use of renewable energies or raw materials.

Why do we need to protect the environment?

There are numerous reasons for protecting the environment, which in turn dictates our own survival. These include:

- **climate change;**
- **depletion of the ozone layer;**
- **acid deposition;**
- **air quality (smog);**
- **water quality;**
- **hazardous substances;**
- **waste;**
- **contaminated land;**
- **population growth.**

As previously stated, environmental pollution can be localised or it can have global impact.

Pollution can be caused through natural disasters which man has little or no control over (e.g. volcanic eruptions, natural forest fires). However, with a population of over 6 billion people, man's actions would be the biggest contributor of environmental pollution today.

To sustain our life and life styles we are consuming more and more raw materials from the Earth and transporting them all over the planet. The process of extraction also contributes to the overall level of pollution. We are also consuming more energy than ever before, primarily using non-renewable fuels (natural gas, coal, oil products), resulting in pollution to air, land and water.

Our output of carbon dioxide from burning fossil fuels is a major cause of climate change, often referred to as global warming or the greenhouse effect. This effect has its own consequences which include an increase in global temperatures, a rise in sea levels and so on.

The release of chlorofluorocarbons (CFCs) primarily from refrigerant gases, fire extinguishers, paints and chemicals has also caused damage to the ozone layer.

While a large part of the world's surface is covered by water, most of this is sea water and is not suitable for drinking. It has to be treated if it is to be used as process water. Our water consumption is growing as we become more industrialised and water is becoming more expensive as our needs for it grow.

Land is also being abused through contamination, waste storage or illegal waste dumping. Forests are being destroyed for their timber or to make use of the land space.

The direct impact on air, land and water is also influencing the range of species of plant, animal and other life forms in the various media. This impact on biodiversity (the variety of life on earth) can destroy ecosystems and the natural balance of life. The main threats to ecosystems include:

- climate change;
- introduced species;
- industrial agriculture and forestry;
- habitat loss and fragmentation;
- genetically modified organisms;
- pollution of land, water and air;
- over-exploitation of plant or other species.

While it is fair to say that the damage already done is unlikely to be reversed, as man's knowledge increases we have the ability to slow down the rate of pollution and in some cases stop it completely if we choose to do so.

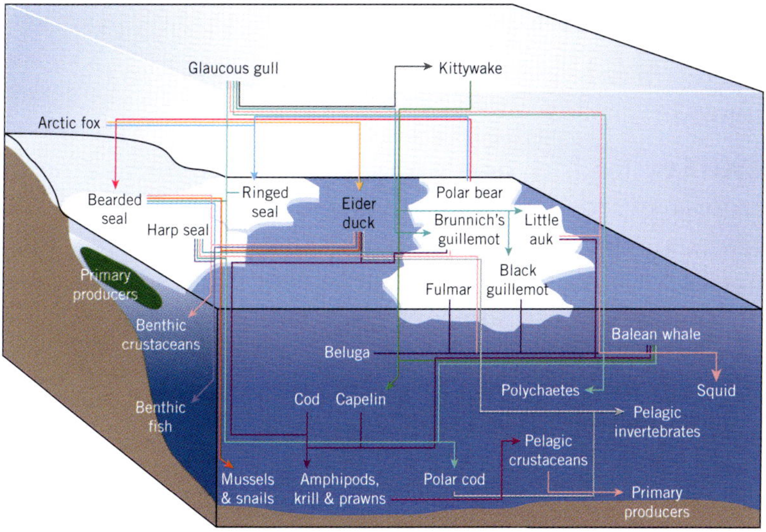

An example of an ecosystem

Q

Can you describe a local ecosystem close to your work or home?

ENVIRONMENTAL LAW

There are two important areas of law relating to the environment.

Civil law tends to deal with the area of compensation or redress awarded or claimed when incidents occur through environmental negligence or nuisance, for example, high noise levels from a factory environment in a residential area or spilling oil onto someone else's land.

Civil court

Criminal law deals with the punishment awarded by the courts when an employer breaks the law, for example, polluting a water course.

Magistrates' court

THE ENVIRONMENTAL PROTECTION ACT

The Environmental Protection Act 1990 (the Act) aims to:

- control pollution of the air, water and land by regulating the management of waste and the control of emissions;
- impose a duty of care on businesses and individuals who produce, carry, keep, treat, dispose of or import controlled waste to do so safely;
- cover statutory nuisances, amend laws relating to litter and place controls on genetically modified organisms and certain other substances.

The act is broken down into parts which deal with specific issues, for example:

- prescribed processes (those deemed to be of high risk to the environment) requiring authorisations to operate;
- the regulation and licensing of controlled waste;
- contaminated land;
- statutory nuisances and litter;
- genetically modified organisms.

LAW ENFORCEMENT

The policing and monitoring of the Environmental Protection Act in the UK is carried out and enforced by a number of agencies including:

- The Environment Agency (EA);
- The Scottish Environmental Protection Agency (SEPA);
- Local authority Environmental Health Officers (EHO).

They have a wide range of powers, which include the power to:

- enter any premises at any reasonable time or, in an emergency, where they believe there is imminent risk of serious pollution to the environment, at any time, and if needed by force;
- take other authorised persons with them;
- take samples of any articles or substances found;
- carry out investigations and examination as necessary;
- dismantle articles or substances which have caused or could cause pollution to the environment;
- inspect and copy records;
- take statements from key personnel;
- require assistance.

Their primary function is to help to ensure employers and others are complying with the relevant regulations and not polluting the environment. In their enforcement role they can issue a:

- **Stop notice** – causing immediate stop of an unlawful activity that is causing harm or has the potential to cause harm to human health or the environment;
- **Compliance notice** – requiring return to compliance within a specified time period;
- **Restoration notice** – requiring a business to make good damage/harm done;
- **Enforcement notice** – requiring compliance with the conditions of a permit;
- **Groundwater prohibition notice** – to prevent an activity that could cause pollution of ground waters;
- **Notice to remove waste;**
- **Anti-pollution works notice.**

They can also:

- revoke, vary or suspend a permit to stop an activity or process to protect the environment;

- prevent or put right (remedy) pollution that has already occurred and recover costs;
- prosecute.

PENALTIES

In a magistrates' court the maximum fine for an environmental offence is **£50,000** and/or a **six-month prison sentence.**

In a crown court the maximum fine is **unlimited** and in addition a **five-year** prison sentence can be imposed.

Q

Do you know who has specific responsibilities for managing environmental issues within your workplace? (If the answer is no then find out.)

EXERCISE 1 – THE ENVIRONMENT AND ENVIRONMENTAL LAW

Self-assessment questions

To assess your level of understanding please complete the following exercise.

1. What is the maximum fine you can receive at a magistrates' court?

   a) £10,000
   b) £50,000
   c) £20,000
   d) £30,000

2. What are the two important types of law?

   a) Civil and matrimonial
   b) Criminal and peaceful
   c) Civil and criminal
   d) Peaceful and matrimonial

3. Can an enforcement officer inspect and copy records?

   a) Yes
   b) No

4. List four reasons why employers should avoid polluting the environment:

   a) ....
   b) ....
   c) ....
   d) ....

5. Give two reasons why employees should take environmental management seriously:

a) ....

b) ....

6. Who are the primary enforcement agencies for environmental law in the UK?

a) Factory Inspectors and firemen
b) Policemen and firemen
c) Environment Agency, Scottish Environmental Protection Agency and local authority Environmental Health Officers (EHOs)
d) Firemen and Environmental Health Officers

7. Environmental law places a duty on:

a) Employers only
b) Employers and employees
c) Employers, employees and the self-employed
d) Nobody

You will find the answers on page 97.

MODULE TWO

Environmental regulations

OIL SPELLS DISASTER IN GULF OF MEXICO FOLLOWING THE DEEPWATER HORIZON BLAST

Welcome to module two. In this module you will learn more about specific Environmental Regulations that are likely to apply to your place of work.

EMISSIONS TO AIR

The air quality standards regulations are designed to limit the concentrations of pollutants that can cause damage through the air. Although man has been polluting the air ever since the first fire was lit, it's our activities since the industrial revolution in the eighteenth and nineteenth centuries that have really had a significant impact.

The main (primary) pollutants entering the air directly from industry and transportation include:

- oxides of carbon including carbon dioxide and carbon monoxide;
- methane;
- oxides of sulphur including sulphur dioxide and sulphur trioxide;
- nitrogen monoxide;
- hydrocarbons (including volatile organic compounds – VOCs);
- suspended particulate matter (SPM);
- some heavy metals, for example lead.

Other (secondary) pollutants (those formed in the air through chemical reactions between primary pollutants and/or other air components) include:

- nitrogen dioxide;
- sulphates;
- nitric acid;
- sulphuric acid;
- hydrogen peroxide;
- ozone (O_3);
- peroxyacetyl nitrate;
- nitrate and sulphate salts.

These pollutants can in turn:

- affect human health, for example respiratory problems;
- affect environmental systems, for example due to accumulation (build up of pollutants) and synergistic effects;
- cause damage to buildings, for example acid rain, ozone, soot and dust;
- contribute to the greenhouse effect (global warming);
- create smog;
- damage the ozone layer;
- return to the ground as acid rain or deposition.

Typical emissions to air

Concern	Key aspects include:
Climate change	• CO_2 • Energy production and use • CFCs
Stratospheric ozone depletion	• CFCs and 1,1,1 tri • Refrigeration and solvent use
Acid rain	• SO_2 and NOx • Combustion including vehicles
Ground-level ozone	• NOx and VOCs • Transport, combustion and solvent use
General air quality	• Hazardous emissions • Particulates

CONTROL OF AIR POLLUTANTS

There are a number of ways employers can control pollution to air. These include:

- elimination at source; this may be achieved by ceasing an activity, changing a process or substitution;
- process control/modification to reduce/avoid the production of pollutants;
- the use of removal technology, for example:
  - collection of particulate matter prior to them becoming airborne;
  - absorption/removal of toxic gases prior to them becoming airborne.

Clearly the use of removal technologies requires not just initial investment but ongoing maintenance and monitoring regimes.

Q

What forms of pollution does your employer discharge to air and how are these controlled?

EMISSIONS TO WATER

Water is fundamental to life both in terms of quantity and quality and although two-thirds of the Earth's surface are covered in water most of it is not suitable for consumption due to its high salt content.

We need to ensure that the water available to us is protected and maintained and the water cycle plays a huge part in this natural process.

The water cycle is the natural process where water evaporated from the sea, freshwater sources, land and that which is transpired by plants is held in the atmosphere, the water condenses into clouds, rain then falls from the clouds providing pure water (unless it is contaminated by impurities in the air). The rain collects on the Earth's surface

The water cycle

(streams, rivers, lakes, reservoirs) or percolates underground to collect in porous rock (aquifers). It is vitally important to prevent pollution of ground water or aquifers.

Pollution to water can be either from specific (point) sources such as:

- discharges from sewage-treatment works to rivers;
- discharges of industrial waste waters to rivers;
- discharges of farm effluents to rivers;
- discharges from small domestic sewage-treatment plants to rivers;
- discharges by means of wells or boreholes into underground strata;
- discharges of collected landfill leachate to rivers;

or from non-point sources such as:

- run-off and under-drainage from agricultural land into rivers;
- general contamination of recharge rainfall to outcropping aquifers;

- septic tank soakaways into permeable strata;
- wash-off litter, dust and dry fallout from urban roads to rivers;
- general entry of sporadic and widespread losses of contaminants to rivers;
- seepage of landfill leachate to underground strata and rivers.

Q

What forms of pollution does your employer discharge to water and how are these controlled?

Water pollution

Pollutants can be dissolved (for example oils, detergents, metal salts and pesticides) or they can be suspended (for example organic matter, inorganic particles and litter).

Each of these pollutants will have an effect on the water in different ways, for example:

- suspended solids can reduce the light entering the water and affect photosynthesis;
- some pollutants raise water temperatures (promoting growth or killing plant/animal life);
- acids or alkalis can affect the pH balance of the water making it toxic to animals/plants;
- plant nutrients such as fertilizers can be toxic to life forms within the water;
- sewage can impact the oxygen availability in the water;
- pesticides can be toxic to life forms within the water;
- pathogens such as bacteria and viruses (for example in raw sewage) place disease directly into the water.

The quality of water can be assessed using a number of key tests, either chemical or biological. These include:

- BOD – Biological Oxygen Demand is a measure of the amount of biodegradable organic matter present;
- COD – Chemical Oxygen Demand is a measure of the absolute oxygen demand by both organic and inorganic matter within a water sample;
- DO – Dissolved Oxygen is a measure of the quantity of oxygen in the sample;
- Biotic Indices look at the diversity and number of different organisms found within a sample of water and each species or group is ranked according to its ability to tolerate pollution;
- pH value (alkalinity or acidity) of the water sample;
- the total suspended solids in the sample;
- the temperature of the sample.

CONTROLLED WATERS

Virtually all fresh and saline natural waters right out to the UK offshore territorial limit are classed as controlled waters and include: rivers and streams, canals, relevant lakes and ponds, estuaries, coastal waters and ground waters. The quality of these waters has to be maintained and this is generally achieved through authorising discharges to controlled waters by a system of consents. The consents issued may impose conditions on the discharger to ensure minimum standards of effluent quality and quantity to maintain the water quality.

DRAINS WITHIN THE WORKPLACE

Generally there are two types of drain within the workplace or home and it is important to recognise them and their purpose:

- Foul water drains are designed to carry contaminated waste water safely to sewage works for treatment.
- Surface water drains are designed to carry uncontaminated rainwater and lead directly to a local river, stream or soak-away.

It is important to ensure that waste water or other contaminants do not enter surface water drains. Many employers will colour code their drains to distinguish between the two.

It is equally important to plan for accidental spillages of oils or chemicals which could enter drains.

Q

Has your employer identified foul and surface water drains and are they easy to recognise?

Pollution going into a drain

A partially blocked drain

A colour coded surface water drain

CONTROL OF WATER POLLUTANTS

Pollution to water systems can be controlled by:

- prevention, that is, eliminating the source of pollution;
- minimising the amount or extent of pollution;
- rendering harmless any pollutant prior to it coming into contact with the water.

EMISSIONS TO LAND

Contamination – the consequence of uncontrolled disposal of waste materials onto land, either deliberately or accidentally.

There are many ways in which land can be contaminated and these are often linked to industries, including:

- coal and mineral mining and preparation;
- smelters, foundries, steel works and metal processing and finishing installations;
- heavy engineering and engineering works, for example car manufacture, ship building;
- electrical and electronic equipment manufacture;
- gasworks, coal carbonisation plants, power stations;
- oil refineries, petroleum storage and distribution sites;
- manufacture of asbestos, cement, lime and gypsum;
- manufacture of organic and inorganic chemicals including pesticides, pharmaceuticals, detergents and cosmetics;
- rubber industry, including tyre manufacture;
- desalination plants;
- munitions producing and testing sites;
- glass making and ceramics manufacture;
- textile industry including tanning;
- paper and pulp manufacture, and printing works;
- timber treatment;
- food processing;
- railway depots, dockyards, garages, road haulage depots, airports;
- landfill and incineration of waste;
- sewage works and farms;
- burial of diseased livestock;
- scrapyards;
- aluminium-production factories;
- dry-cleaning premises.

Contaminated land

Potential land pollution

Routes of contamination

The routes of contamination vary and can include:

- leakages from underground pipes or tanks;
- leakage during storage or transport of materials above ground;
- accidental spillages;
- burial or tipping of waste onto the site (letting it run away into the ground or evaporate: this may lead to the formation of "hot-spots" on an otherwise clean site);
- prolonged deposition from airborne emissions (settling back to earth from an industrial vent or stack along the line of the prevailing wind);
- careless demolition, releasing harmful substances still stored on site (e.g. tank contents or asbestos building materials);
- movement of contaminated ground water from adjacent areas;
- migration of toxic or explosive gases from adjacent areas.

Pollutant linkage

When pollution has been identified it is important to identify the pathway and the source in order to prevent further pollution. Conversely employers can identify their sources of pollution and the potential pathways they may take to a known receptor.

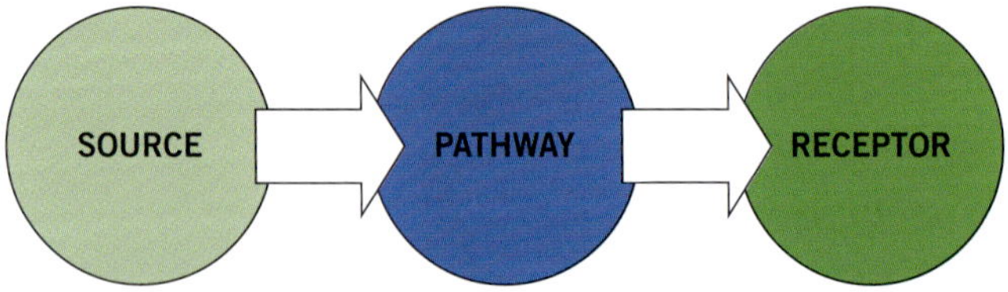

A typical contamination route

A pollution source

A pollution pathway

A pollution receptor

Q

What forms of pollution does your employer discharge to land and how are these controlled?

CONTROL OF LAND POLLUTANTS

Pollution to land can be controlled by:

- prevention via elimination or substitution;
- containing the contaminants on site, isolating them to avoid further spread;
- removing the contaminated material from the site and transporting it to a secure, licensed disposal site;
- treating it in-situ to permanently remove/destroy the contaminants. Treatment techniques include:
    - washing contaminated soil;
    - bioremediation (using biological agents);
    - soil vapour extraction;
    - thermal stripping;
    - incineration;
    - vitrification;
    - solidification;
    - chemical treatment.

STORAGE OF MATERIALS

It is important to ensure that all materials are stored correctly in appropriate containers and in the correct locations to prevent potential land pollution, fires or dispersal.

Care should be taken to prevent liquid waste or leachate entering drains or ground waters from storage areas.

Poor waste material storage

SPILLAGE CONTROL

Procedures should be in place to attempt to prevent spillages or at least control them and minimise any impact if they do occur. Any controls introduced should also consider the potential for fire water runoff.

WASTE

Waste can be defined as "any substance which constitutes a scrap material or an effluent or other unwanted surplus substance arising from a process and any substance or article which requires disposal due to it being broken, worn-out, contaminated or otherwise spoiled".

Waste can be further categorised as:

- **DIRECTIVE WASTE**, which is any substance or object discarded or intended to be discarded;
- **CONTROLLED WASTE**, which is subject to the Duty of Care under the EPA Part II and includes industrial, commercial and household waste;
- **HAZARDOUS WASTE**, which may be toxic, harmful, irritant, corrosive and so on.

The current problem with waste in the UK is that we are producing too much of it and are running out of places to dispose of it economically.

Historically landfill has been the preferred option. However, there are several issues with landfill which include:

- finite usable disposal space;
- risk of land contamination;
- risk of water contamination (leachate);
- landfill gas is explosive and contributes to the greenhouse effect;
- related nuisances such as vermin, noise or odours;
- excessive waste transportation distances adding to the pollution effect.

To combat the environmental problems associated with excessive travel distances (sometimes as far as other countries) the proximity principle has been adopted, which states that any waste generated should be disposed of as near as possible to its point of production.

Waste

Q

What are the main wastes generated by your employer and how are they disposed of?

CONTROL OF WASTE

There are several options available to manage waste based on a hierarchy, and these include:

- waste reduction
    - the use of less material in the product;
    - reducing the amount of waste from the manufacturing process;
    - extending the life of the products;
- re-use
    - greater use of returnable containers;
    - re-usable transit packaging;

- recovery
  - recycling;
  - composting;
- recovering energy
  - creating and using energy from burning waste;
  - using landfill gases;
- proper disposal
  - incineration or landfill.

Employers can assess the amount of waste they are handling by carrying out a waste audit, which includes identifying all waste inputs and outputs through their organisation.

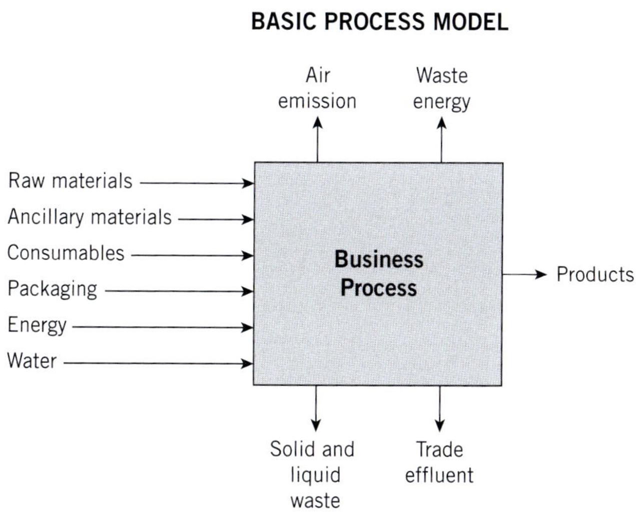

Process model

WASTE LEGISLATION

Generally it is an offence to:

- deposit controlled waste in or on land unless a waste management licence is in force;
- treat, keep or dispose of controlled waste unless done so in accordance with a waste-management licence;
- treat, keep or dispose of controlled waste in a manner likely to cause pollution of the environment or harm to human health.

The duty of care principle:

It is the duty of ANY person who imports, produces, carries, keeps, treats, disposes of or acts as a broker of controlled waste to take reasonable measures to:

- prevent contraventions, by any other persons;
- prevent the escape of waste;
- transfer waste only to authorised persons;
- ensure a written description of the waste is provided;
- contain it securely to avoid escape;
- transfer it only to someone with the authority to take it (e.g. registered or exempt carrier/licensed waste manager);
- keep appropriate records using transfer notes (including description and information on any potential problems).

Note: Householders are exempt for their own household waste from the above duties. However, it is still important to manage your own domestic waste as best you can in accordance with local authority rules.

The receivers of waste MUST:

- have the AUTHORITY TO RECEIVE IT;
- ENSURE the documentation is properly completed.

Note: All waste holders should be alert to the non-compliance of others.

Waste management

It is an offence to treat, keep or dispose of waste without a Waste Management Licence, unless the operation is excluded or exempt from requiring a licence.

WASTE-DISPOSAL OPTIONS

There are several options, including:

- tipping of waste above or underground (illegal fly tipping of waste is becoming an increasing problem due to the increasing costs of waste disposal);
- land treatment of waste;
- deep injection of waste;
- surface impoundment of waste;
- specially engineered landfill of waste;
- release of solid waste into a water body – except seas or oceans;
- release of solid waste into seas or oceans including seabed insertion;
- biological treatment of waste not listed elsewhere;
- physico-chemical treatment of waste not listed elsewhere;
- incineration of waste on land;
- incineration of waste at sea;
- permanent storage of waste (e.g. radioactive or asbestos waste);

- blending or mixture of waste prior to it being submitted to any of the operations listed above;
- repackaging of waste prior to the waste being submitted to any of the operations listed above;
- storage of waste prior to it being submitted to any of the operations listed above (excluding temporary storage, pending collection, on the site where the waste is produced).

Waste recovery options

Waste recovery is becoming increasingly popular as raw materials become scarcer, and disposal options become more difficult or expensive. Options for waste recovery include:

- reclamation or regeneration of solvents;
- recycling or reclamation of organic substances not used as solvents;
- recycling or reclamation of metals and metal compounds;
- recycling or reclamation of other inorganic compounds;
- regeneration of acids or bases;
- recovery of components used for pollution abatement;
- recovery of components from catalysts;
- re-refining, or other re-uses, of waste oil;
- use of waste principally as a fuel or for other means of generating energy;
- spreading of waste on land resulting in benefit to agriculture or ecological improvement, composting, other biological transformation processes unless otherwise excluded;
- use of waste obtained from any of the operations listed above;
- exchange of wastes for submission to any of the operations listed above;

- storage of waste consisting of materials intended for submission to any of the operations listed above, excluding temporary storage, pending collection, on the site where it is produced.

Hazardous wastes

Hazardous wastes include:

- acids;
- alkaline solutions;
- batteries;
- oils;
- industrial solvents;
- pesticides;
- wood preservatives;
- pharmaceutical compounds;
- fluorescent tubes & TVs;
- refrigerators.

An example of hazardous waste

Hazardous wastes require additional precautions when being transported. This includes the use of **consignment notes** (documentation designed to identify waste, its ownership from producer to accepter and to track the waste through to final disposal). The consignment note requires the following to be stipulated:

1. The consignor must provide:
   - details of the premises the waste is being removed from;
   - the waste's destination;
   - the waste producer;
   - the premises code;
   - consignment note code (premises code and unique number);
   - a description of the waste;
   - the European waste catalogue number.
2. The carrier (person transporting the waste) must provide proof of registration or exemption and certify that the details provided by the consignor are correct.
3. The consignee (person recovering or disposing of the waste) confirms acceptance of the waste as described.

Each party signs the form and keeps a copy on record for a minimum of three years (12 months for the carrier).

Consignee

The consignee is required to:

- Notify the consignor that the waste has been accepted (e.g. a copy of the note).
- Send the Environment Agency (EA) a quarterly summary of the wastes accepted or rejected.

- Pay a fee to the EA for each consigned load. The consignee can charge the consignor this fee.

Producer responsibility

Producers of waste have a duty to ensure that the products and materials they produce are put to productive use when they have served their original purpose. Target waste streams include:

- newspaper;
- packaging waste;
- consumer and automotive batteries;
- tyres;
- vehicles;
- electronic equipment.

Waste electrical and electronic equipment

By making producers responsible for dealing with the products at their "end of life", there is a greater incentive for them to improve product design to facilitate recycling. Producers also have to contribute to the cost of any waste collection schemes, and pay for re-use or recycling of their products as they become waste.

Vehicles – end of life

Within the motor industry, manufacturers are improving the re-use, recycling and other forms of recovery of end-of-life vehicles and their components, with a view to reducing waste disposal. The last holder and/or owner has the right to deliver vehicles to treatment facilities, without having to bear any cost.

Waste car parts

ENVIRONMENTAL NUISANCES

In addition to environmental pollution there are a number of situations that, when prejudicial to human health, are classed as environmental nuisances. These include:

- premises in a poor state (e.g. rubbish accumulation, excessive numbers of pets);
- smoke emitted from premises;
- fumes or gases from private dwellings;
- dust, steam, smell or other effluvia from industrial, trade or business premises;
- any accumulation or deposit;
- any animal kept in such a place or manner as to be prejudicial to health or a nuisance;
- noise emitted from premises;
- noise from a vehicle, machinery or equipment in the street.

Environmental nuisances are normally dealt with by local authority Environmental Health Officers.

Q

Is your employer causing any environmental nuisances, and if so what could be done to stop them?

EXERCISE 2 – ENVIRONMENTAL REGULATIONS

Self-assessment questions

To assess your level of understanding please complete the following exercise.

1. List three examples of typical air pollutants:

   a) ....

   b) ....

   c) ....

2. Briefly describe two effects of global warming:

   a) ....

   b) ....

3. List two examples of tests that can be carried out to assess water quality:

   a) ....

   b) ....

4. Add the missing word to the following sequence: source, pathway, xxxxxxxx:

   a) ....

5. What are the two types of drain typically found at most workplaces and homes?

   a) ....

   b) ....

6. List three waste streams targeted under producer responsibility:

   a) ....

   b) ....

   c) ....

7. List two examples of environmental nuisance:

   a) ....

   b) ....

You will find the answers on pages 98–9.

MODULE THREE

Energy

WIND FARMS TURNING HEADS ALL OVER THE COUNTRY

Welcome to module three. In this module you will learn more about the traditional forms of energy and the need to conserve them along with initiatives to harness new renewable energies.

What is energy? Although there are several definitions of energy we will use the following one:

Energy is defined as "the capacity to do work". The energy used may be from manpower, horsepower, fuels, electricity and so on.

$$\text{Energy used} = \text{power applied} \times \text{time in use}$$

We all need energy to perform basic functions on a daily basis. Energy is taken for granted by most, but traditional non-renewable energy supplies are being depleted and their conservation is becoming more and more important. At the same time greater use is being made of renewable energies along with research to help ensure a continual supply of energy for the twenty-first century and beyond.

	Examples of energy usage
Buildings	Heating, cooling, hot water, lighting, catering and appliances
Factories/Warehouses	Motive power, compressed air, cooling, refrigeration, heating and pumping
Process heat	Furnaces, ovens, kilns, boilers, tanks, driers
Transport	Vehicle fuel/power

TYPES OF ENERGY

We obtain energy from many sources. These can include:

- **PRIMARY ENERGY SOURCES** which include fossil fuels (oil, gas, coal);
- **DELIVERED ENERGY** which is the energy that is delivered to the point of use;
- **USEFUL ENERGY** which is the energy in a form required by end user.

There is an obvious link between energy and sustainability and we need to take a more responsible attitude to the use of energy, especially from non-renewable sources.

ENERGY SERVICE	Function it performs
ENERGY CONVERSION	Process of conversion
SECONDARY ENERGY	Energy resulting from conversion
NON-RENEWABLE ENERGY	Created over time and about 98% of energy used cannot be replaced over similar time
RENEWABLE ENERGY	Can be replaced over a human lifetime

Primary energy sources

We have historically relied on the extraction and use of "original" sources of energy from non-renewable fossil fuels (coal, oil, gas). In the twentieth century nuclear power was introduced in developed counties to help supply an ever-growing energy demand. However, we have to recognise that there are safety- and health-related issues with the operating of nuclear power stations and the storage of nuclear waste.

A greater focus is now placed on the use of renewable energies, which include:

- hydro power;
- solar power;
- wind power;
- wave / tidal power;
- geo-thermal power;
- biomass (organic materials from which biofuels can be obtained).

Delivered energy

This is the energy supplied to premises (commercial and residential) in a usable form:

- solid fuels (e.g. coal, coke);
- oil based fuels (e.g. diesel, petrol, fuel oil);
- natural gas;
- electricity.

ENERGY AND ENVIRONMENTAL ISSUES

There are several environmental issues related to providing and distributing delivered energy. These include:

- local impacts, for example visual intrusion;
- pollution incidents, for example oil spills;
- atmospheric releases causing problems at local, regional and global scales;
- natural resource depletion.

Q

What are the impacts in your local area of energy production and supply?

One of the biggest impacts of energy production and distribution is the release of carbon dioxide into the atmosphere and the impact this has on global warming (other gases such as methane also contribute to global warming). In the UK, 97% of emissions of carbon dioxide come from fossil fuel combustion (22% industry, 30% power stations and 22% transport). Carbon dioxide is a natural product of hydrocarbon combustion and there is no current feasible end-of-pipe abatement technology available to prevent it escaping into the atmosphere.

The greenhouse effect

The greenhouse effect/global warming affects us all in different ways:

- rising sea levels as a result of melting ice caps, permafrosts and glaciers and thermal expansion, causing millions of people to be displaced world-wide due to the flooding of low-lying areas;
- changing weather patterns – we are experiencing extreme weather, for example heat waves, cold spells, droughts and floods;
- loss of biodiversity due to the destruction of ecosystems and habitats;
- the worldwide impact on tourism (e.g. loss of ski resorts);

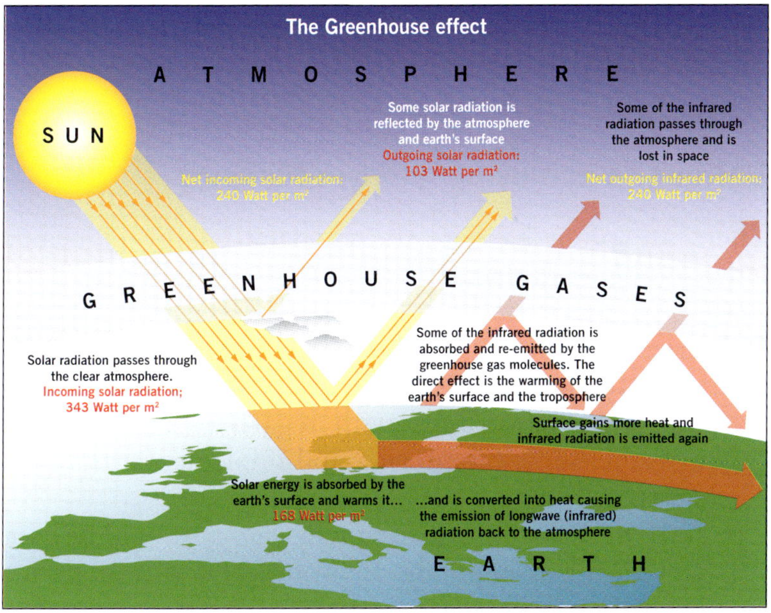

The greenhouse effect

- infrastructure at risk (rainwater affecting buildings and roads in the Middle East where historically rainfall was very low).

There are a number of national and international initiatives to attempt to reduce the amount of carbon dioxide entering the environment.

1. **Switch to non- or low-carbon energy sources, for example**

    - solar sources;
    - low carbon fuels: natural gas releases less CO_2 than oil, oil releases less than coal, and grid electricity is also associated with high CO_2 emissions.

2. **Improve energy efficiency**

    - Use less energy for desired outcome, for example, adjusting comfort levels, production levels.

Barriers to achieving energy efficiency

There are often barriers to achieving energy efficiency either in the workplace or home. These include:

- the perception that energy-saving initiatives are expensive and have a long payback period;
- staff time constraints with regard to carrying out investigations, monitoring and implementation measures;
- staff not being informed of why energy should be conserved, for example the environmental and economic benefits;
- lack of training about how energy can be conserved;

- failure to implement and manage initiatives systematically.

Energy management strategy

In order to achieve energy efficiency within the workplace or home it is important to start with a strategy. A typical strategy would include the following:

- gaining commitment at all levels;
- ensuring everyone understands what it is they are trying to achieve and how it can be achieved;
- organising and prioritising resources;
- deciding how the strategy will be implemented;
- establishing controls and agreeing a monitoring process.

SOLAR POWER

Solar power effectively converts the energy from the sun into a form that can be used. The solar energy falling on the Earth in an hour equals the total energy used by the world in a YEAR. However, its availability, low energy density and collection remain the main obstacles to greater utilisation. Examples of solar energy usage include:

- **passive solar heating:** rays are captured directly through large south-facing windows;
- **active solar heating:** special flat plate or parabolic collectors are used;
- **photovoltaic devices:** use silicon to convert sunlight into current.

Solar panels

WIND POWER (ONSHORE AND OFF SHORE)

When air moves causing wind, it has kinetic energy (the energy created whenever mass is in motion). Using the right technologies, the wind's kinetic energy can be harnessed and converted to other forms of energy such as electricity or mechanical power. This technique was first used in windmills, using wind power to pump water or grind grain.

Today's utility-connected wind turbines and multi-turbine wind farms use wind power to generate clean, renewable energy to power homes and businesses.

Wind turbines

Wind-power generation uses a natural and virtually inexhaustible source of power to produce electricity. Although wind-power generation is clean (it doesn't cause air, soil or water pollution) the building and establishing of wind farms do have environmental impacts.

TIDAL POWER (INCLUDING WAVE POWER)

A daily naturally occurring event associated with oceans is tidal movement. This event moves incredible amounts of water. The process of harnessing energy from the ocean's tide is achieved via turbines. The movement and flow of the tide in and out from the shore is used to turn the turbines and create energy. The turbines' energy is then routed to storage generators and sent out through power lines to feed a supply grid.

As tidal movements are naturally occurring, they are a predictable source of energy. However, as the tide is only moving for 8 to 10 hours per day the power generation is limited, so alternative power stations are used to ensure supply during times when the tidal power is not available. Although tidal power is considered to be free and creates no harmful byproducts that can damage the environment, the building and establishing of the turbines and infrastructure do have environmental impacts.

Water turbine

FRACKING

In addition to looking at renewable energies, governments and corporations are always seeking to identify new techniques to extract what is left of our non-renewable energy sources, and an example of this includes the process of hydraulic fracturing (commonly known as fracking) – a technique designed to recover gas and oil from shale rock.

Fracking involves drilling into the earth and then directing a high-pressure water mixture at the rock to release the gas inside. Water, sand and chemicals are injected into the rock at high pressure which allows the gas to flow out to the head of the well.

The process is carried out vertically or by drilling horizontally to the rock layer. The process can create new pathways to release gas or can be used to extend existing channels.

Although this technique releases as yet untapped reserves of gas, there are a number of environmental concerns, including:

- the fact that huge amounts of water have to be transported to the fracking site, at significant environmental cost;
- potentially carcinogenic chemicals used may escape and contaminate groundwater around the fracking site;
- the fracking process can cause small earth tremors (two small earthquakes of 1.5 and 2.2 magnitude hit the Blackpool area in 2011 following fracking).

TRANSPORT

One of the biggest contributors to environmental pollution in developed countries is the vehicles used to transport goods and passengers.

ASPECT	IMPACT
Consumption of fuel	Resource depletion
Emissions, e.g.:	Local and global air pollution:
• the primary greenhouse gas – CO_2 • CO, VOCs, NOx and particulates	• global warming/climate change • air quality (human health), ground level ozone, acid rain, smog, deposition (NOx)

ASPECT	IMPACT
Noise, vibration and visual intrusion	Nuisance
Discarding of spent materials, e.g. tyres	Waste generation and oil and other materials disposal
High traffic volumes leading to congestion	Enhanced levels of other aspects, e.g. emission levels

Eco friendly vehicle

The following factors will impact the levels of pollution:

- fuel used:
  - petrol: emits more CO_2 and VOCs and less efficient than diesel;
  - diesel: emits more NOx and PM and less CO_2;
- engine design and other pollution control devices (e.g. catalytic convertor);
- vehicle maintenance;
- driver behaviour/traffic conditions:

  - stop/start;
  - quick/slow;
  - loaded/un-loaded and so on;

- cold starts:

  - the colder the start, the more the initial emissions until normal running temperature is achieved.

Some simple measures that can be adopted to improve vehicle efficiency and/or reduce pollution include:

- the purchase of fuel-efficient vehicles;
- correct vehicle maintenance;
- the use of public transport;
- effective route planning;
- driver training;
- employee initiatives, for example cycle to work, car-share schemes;
- the use of alternative fuels, for example LPG, electric, biofuels, hybrid vehicles;
- recruiting locally to avoid long commutes.

Air transport

- By 2050, passenger-kilometres flown world-wide could grow to between five and nine times that in the mid-90s.
- Aircraft produce large amounts of toxic emissions that are a threat to human health, including nitrogen oxides and volatile organic compounds (VOCs).
- It is estimated that 10% of climate change will be directly attributable to aircraft by 2050.

Although technology is being used to make aircraft travel more efficient, this does not make a significant impact on overall emissions.

Passenger taxes have also been introduced. However, these do not directly impact aircraft emissions.

EXERCISE 3

Self-assessment questions

To assess your level of understanding please complete the following exercise.

1. What are the three fossil fuel, non-renewable energies we have traditionally relied on?

   a) ....

   b) ....

   c) ....

2. List three renewable energies that are being utilised:

   a) ....

   b) ....

   c) ....

3. Briefly provide two reasons why we can no longer rely on fossil fuels:

   a) ....

   ....

   b) ....

   ....

4. Define the term energy:

a) ....

....

....

....

5. List three different types of pollution from vehicle usage:

a) ....

b) ....

c) ....

6. Give four examples of reducing the levels of pollution associated with vehicle usage:

a) ....

b) ....

c) ....

d) ....

You will find the answers on pages 99–100.

MODULE FOUR

Proactive management of the environment

THE PLANTING OF JATROPHA TREES IN DESERT AREAS COULD BE AN EFFECTIVE WAY OF CURBING EMISSIONS OF CARBON DIOXIDE

Welcome to module four. In this module you will learn more about some of the practical steps you and your employer can take to help identify the environmental aspects resulting from activities taking place and introduce measures to reduce/control the environmental impacts of the operations to minimise environmental pollution. Remember your employer can only succeed with your help, co-operation and commitment.

Man has tried to manage his environment since the dawn of time. However, as the world population has grown and our understanding of the impacts has improved, the effects of pollution have become more acute and apparent. We now have a greater appreciation of how actions taken in one continent can impact other continents thousands of miles away directly and indirectly.

More and more employers, organisations and individuals are taking a far more responsible attitude to environmental management and there are several management system models available to help employers achieve this.

EMAS – ECO-MANAGEMENT AND AUDIT SCHEME

The EMAS system was developed by the European Community (EU) and is recognised throughout the EU. It requires the production of public environmental statements.

The EMAS system consists of the following elements:

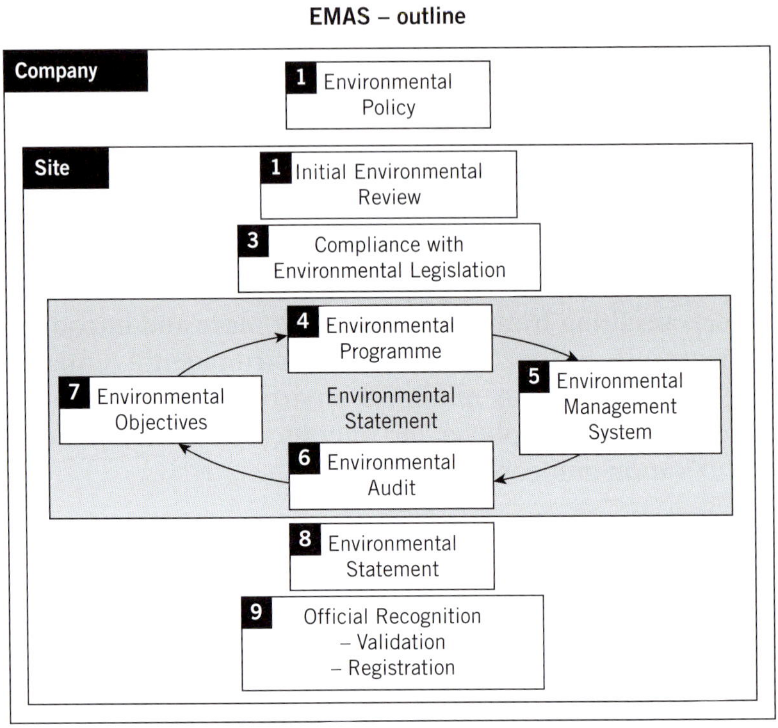

EMAS model

ISO 14001 – ENVIRONMENTAL MANAGEMENT SYSTEM

This system is popular in the UK and was developed by the International Standards Organisation (ISO). As such it is a recognised International Standard and follows a similar structure to that used within ISO 9001 (the Quality Standard).

There are several incentives for employers to introduce an EMS and these include:

- greater awareness of relevant legislation and compliance;
- appeasing environmental campaign groups;
- media pressure;
- maintaining/improving public image;
- keeping up with competitors;
- appeasing neighbours;
- increasing customer base;
- impact on insurance premium;
- appeasing investors.

The ISO 14001 management system consists of the following elements:

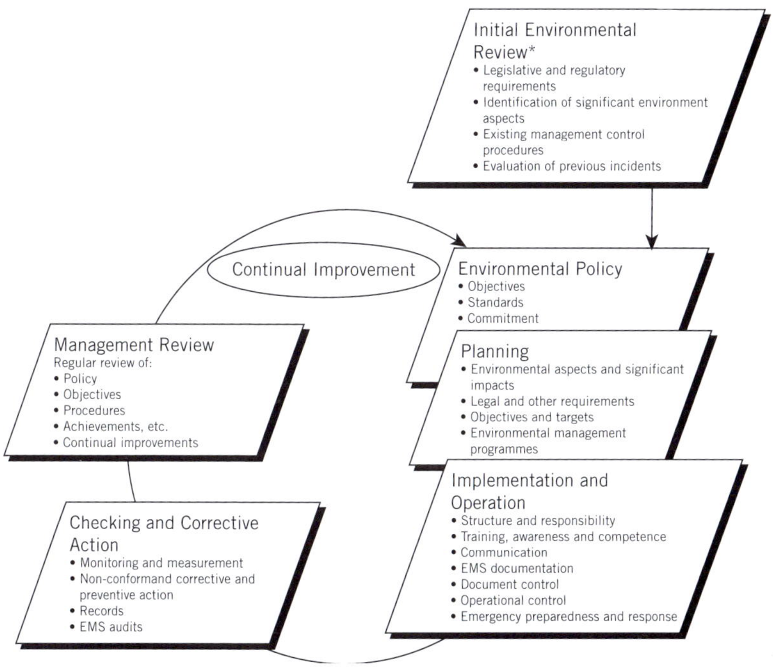

The 14001 model

Within the ISO 14001 management system there are some key definitions, which include:

The environment

Surroundings in which an organisation operates including air, water, land, natural resources, flora, fauna, humans, and their interrelation.

Note: Surroundings in this context extend from within an organisation to the global system.

Environmental aspect

These are the elements of an organisation's activities, products or services that can **interact** with the environment.

Note: A significant environmental aspect is an environmental aspect that has or can have a significant environmental impact.

Environmental impact

Any **change** to the environment, whether **adverse or beneficial**, wholly or partially resulting from an organisation's activities, products or services.

Any organisation undertaking an environmental review will carry out a review of their aspects and impacts by following the steps below:

Step 1 – Select an activity, product or service

Step 2 – Identify environmental aspects

- to air, land and water;
- usage of materials and resources;
- visual and nuisance;
- consider normal, abnormal and emergency situations.

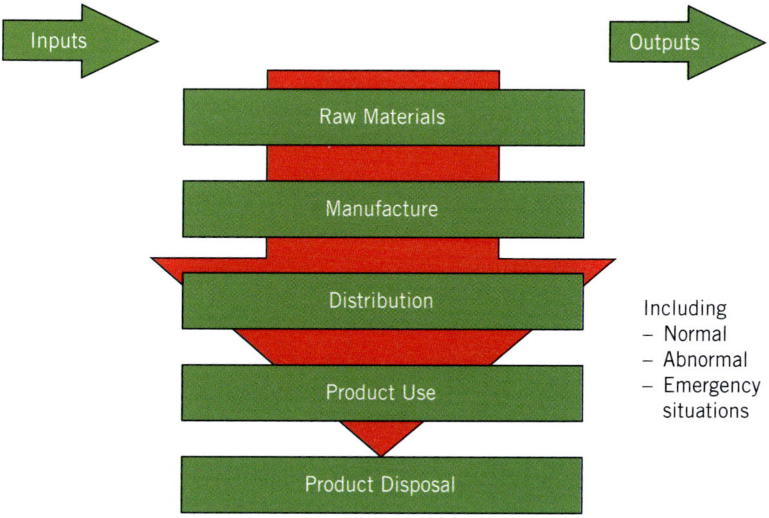

Typical industrial inputs and outputs

Step 3 – Identify environmental impacts

- direct/indirect;
- local/global;
- positive/negative.

Step 4 – Evaluate significance of impacts with respect to the environment and business

When evaluating the significance of the impacts it is important to consider both environmental and business concerns, taking into account local needs and sentiment.

Environmental Concerns	Business Concerns
Scale	Legislation
Severity	Stakeholders/public image
Probability	Cost of controls
Duration	Practicability of controls

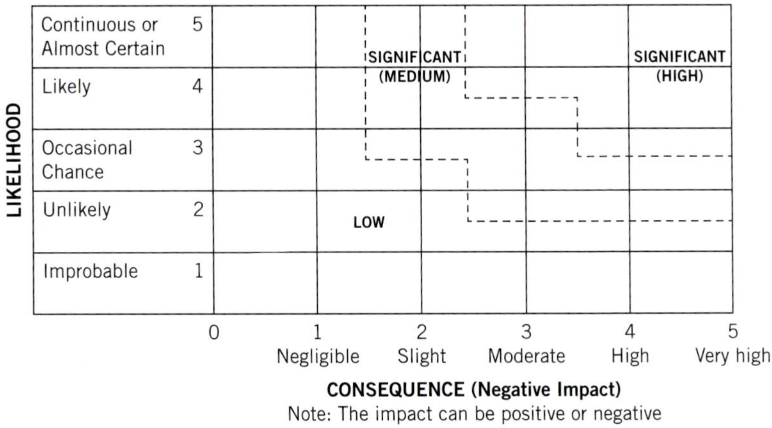

Priority indicator

LIFE-CYCLE ASSESSMENT (LCA)

More and more employers are taking a different view of their operations and carrying out Life-cycle Assessments (LCAs, also known as life-cycle analysis, eco-balance and cradle-to-grave analysis). This is a technique used to assess the environmental impacts associated with all the stages of a product's life from cradle to grave (i.e., from raw material extraction through materials processing, manufacture, distribution, use, repair and maintenance, and disposal or recycling).

LCAs can help employers take a more holistic approach to their operations and the impacts they have on the environment. The key elements of the process include:

- compiling an inventory of relevant energy and material inputs and environmental outputs;
- evaluating the potential impacts associated with identified inputs and outputs;
- interpreting the results to help make a more informed decision about what they do.

ECOLOGICAL (OR ECO) FOOTPRINT

By studying the **ecological footprint** of what we do or consume we can measure the human demand on the Earth's ecosystems. This is a standardised measure of demands placed on the natural resources that are available compared to the planet's ecological capacity to regenerate what is consumed. Eco footprints represent the amount of biologically productive land and sea area necessary to supply the resources the human population consumes (including the associated waste assimilated). Using this assessment, it is possible to estimate how much of the Earth (or how many planet Earths) it would take to support humanity if everybody followed a given lifestyle. In 2007, it was estimated that humanity's total ecological footprint was 1.5 planet Earths, that is, humanity uses ecological services 1.5 times more quickly than the Earth can renew them. Using this system we can estimate the eco footprint of our activities or the products we buy or use.

PLANNING FOR ENVIRONMENTAL EMERGENCIES

Any business or organisation that could potentially cause pollution to the environment through its activities should develop a plan to deal with foreseeable events. In addition it should have available the equipment and resources necessary to deal with the identified events.

Consideration should also be given to carrying out simulated exercises to mimic the reaction to adverse events and help to train key employees. If possible external agencies should be included in the exercise process.

Waste run off from a fire

EXERCISE 4

Self-assessment questions

To assess your level of understanding please complete the following exercise.

1. ISO 18001 is the environmental management system

   a) True
   b) False

2. Define the term environmental aspect:

....

....

....

3. Define the term environmental impact:

....

....

....

4. List three examples of environmental concerns an organisation may have when evaluating the significance of impacts:

a) ....

b) ....

c) ....

5. List three examples of business concerns an organisation may have when evaluating the significance of impacts:

a) ....

b) ....

c) ....

6. Does the ISO environmental management system require continuous improvement?

a) Yes
b) No

7. List three reasons why employers should plan for emergencies:

a) ....

b) ....

c) ....

You will find the answers on pages 100–1.

Answers to questions

MODULE ONE – The environment and environmental law

1. What is the maximum fine you can receive at a magistrates' court?

   b) £50,000

2. What are the two important types of law?

   c) Civil and criminal

3. Can an enforcement officer inspect and copy records?

   a) Yes

4. List four reasons why employers should avoid polluting the environment.

   Legal, moral, financial, reputational (examples of these are acceptable)

5. Give two reasons why employees should take environmental management seriously.

   It affects them and their families; it can save them money; compliance with employer policies; social pressure

6. Who are the primary enforcement agencies for environmental law in the UK?

   c) Environment Agency, Scottish Environmental Protection Agency and local authority Environmental Health Officers (EHOs)

7. Environmental law places a duty on:

   c) Employers, employees and the self-employed

MODULE TWO – Environmental regulations

1. List three examples of typical air pollutants.

   Carbon dioxide, carbon monoxide, methane, sulphur dioxide, sulphur trioxide, nitrogen monoxide, volatile organic compounds, suspended particulate matter, heavy metals, nitrogen dioxide, sulphates, nitric acid, sulphuric acid, hydrogen peroxide, ozone, peroxyacetyl nitrate

2. Briefly describe two effects of global warming.

   Rising sea levels, population displacement due to flooding, changing weather patterns, loss of biodiversity, destruction of habitats, impact on tourism, damage to infrastructure

3. List two examples of tests that can be carried out to assess water quality.

   Biological Oxygen Demand, Chemical Oxygen Demand, Dissolved Oxygen, organism diversity (Biotic Indices), pH value (alkalinity or acidity), total suspended solids, water temperature

4. Add the missing word to the following sequence: source, pathway, xxxxxxxx.

   receptor

5. What are the two types of drain typically found at most workplaces and homes?

   Surface water, foul water

6. List three waste streams targeted under producer responsibility.

   Newspaper, packaging waste, consumer and automotive batteries, tyres, vehicles, electronic equipment

7. List two examples of environmental nuisance.

Premises in a poor state; smoke emitted from premises; fumes or gases from private dwellings; dust, steam, smell or other effluvia from industrial, trade or business premises; accumulation of deposits; animals being kept in such a place or manner as to be prejudicial to health or a nuisance; noise emitted from premises; noise from a vehicle, machinery or equipment in the street

MODULE THREE – Energy

1. What are the three fossil fuel, non-renewable energies we have traditionally relied on?

Oil, gas, coal

2. List three renewable energies that are being utilised.

Hydro power, solar power, wind power, wave/tidal power, geothermal power, biomass

3. Briefly provide two reasons why we can no longer rely on fossil fuels.

Limited supply remaining, cost of extraction, impact on environment, health and safety of employees, hinders research into new technologies

4. Define the term energy.

The capacity to do work

5. List three different types of pollution from vehicle usage.

Exhaust emissions (carbon dioxide and monoxide, VOCs, particulates), noise, visual impact, discarded materials (tyres, oil, batteries etc.)

6. Give four examples of reducing the levels of pollution associated with vehicle usage.

   The purchase fuel efficient vehicles, vehicle maintenance, use of public transport, effective route planning, driver training, walk or cycle to work, car-share schemes, use of alternative fuels, recruiting locally to avoid long commutes

MODULE FOUR – Proactive management of the environment

1. ISO 18001 is the environmental management system.

   b) False

2. Define the term environmental aspect.

   The elements of an organisation's activities, products or services that can interact with the environment

3. Define the term environmental impact.

   Any change to the environment, whether adverse or beneficial, wholly or partially resulting from an organisation's activities, products or services

4. List three examples of environmental concerns an organisation may have when evaluating the significance of impacts.

   Scale, severity, probability, duration

5. List three examples of business concerns an organisation may have when evaluating the significance of impacts.

   Legislation, stakeholders/public image, cost of controls, practicability of controls

6. Does the ISO environmental management system require continuous improvement?

   a) Yes

7. List three reasons why employers should plan for emergencies.

   Sensible proactive management, cost effective, reputational protection, speed of reaction, preparedness

Glossary/definitions

Accident An unplanned and/or uncontrolled event which has led to injury, damage or other loss to the business

Aquifer An underground water supply found in porous rock, sand or gravel

BAT Best available technology

BOD Biological Oxygen Demand

BPEO Best practicable environmental option

CFC Chlorofluorocarbon

COD Chemical Oxygen Demand

COMAH Control of major accident hazards

COSHH Control of Substances Hazardous to Health regulations

DEFRA Department of Environment Food and Rural Affairs

DO Disolved Oxygen

EA Environment Agency

EC European Commission

EEPA European Environment Protection Agency

EHO Environmental Health Officer

EIA Environmental impact assessment

EMAS Eco-management and audit scheme

EMS Environmental management scheme

EPA Environmental Protection Act

Hazard Something that has the potential to cause harm

H&S Health and Safety

HSE Health and Safety Executive

HSWA Health and Safety at Work Act

IPPC Integrated Pollution Protection Control

LCA Life-cycle Assessment

Leachate Any liquid that, in passing through matter, extracts solutes, suspended solids or any other component of the material through which it has passed

LEV Local exhaust ventilation

MSDS Material Safety Data Sheet

NOx Oxides of nitrogen

PPE Personal Protective Equipment

Risk The probability and consequences of a hazard occurring

RPE Respiratory Protective Equipment

SEPA Scottish Environmental Protection Agency

SOx Oxides of sulphur

SPM Suspended particulate matter

TOD Total Oxygen Demand

VOC Volatile Organic Compound

Examination

Once you have completed the guide you can choose to complete the examination.

Complete the following exam to test your environmental management knowledge. All candidates will be informed whether they have passed or failed via email, and candidates achieving a pass mark of 75% or above will receive a certificate.

To find out details of where to send your completed exam, or to request an electronic version, please visit www.routledge.com/9780415835411

You have thirty minutes – good luck

<table>
<tr><td>Your FULL Name: (PRINT)</td><td></td><td>Employee Reference:</td><td>Employer Name:</td></tr>
<tr><td>Date of Birth:</td><td></td><td colspan="2" rowspan="2">Your Email Address:</td></tr>
<tr><td>Date of Examination:</td><td></td></tr>
</table>

Please indicate your answer(s) by putting a tick in the appropriate box(es)

1. What is the maximum fine you can receive at a magistrates' court?

   a) £10,000 ❐
   b) £20,000 ❐
   c) £30,000 ❐
   d) £50,000 ❐

2. What are the two important types of law?

   a) Civil and matrimonial ❐
   b) Criminal and peaceful ❐
   c) Civil and criminal ❐
   d) Peaceful and matrimonial ❐

3. Can an enforcement officer inspect and copy records?

   a) Yes ❐
   b) No ❐

4. Who are the primary enforcement agencies for environmental law in the UK?

   a) Factory Inspectors and firemen ❐
   b) Policemen and firemen ❐
   c) Environment Agency, Scottish Environmental Protection Agency and local authority Environmental Health Officers (EHOs) ❐
   d) Firemen and Environmental Health Officers ❐

5. Who does environmental law place a duty on?

   a) Employers only ❐
   b) Employers and employees ❐
   c) Employers, employees and the self-employed ❐
   d) Nobody ❐

6. Which of the following are typical air pollutants?

   a) Carbon dioxide gas ❐
   b) Plastic carrier bags ❐
   c) Batteries ❐
   d) Particulate matter ❐

7. Which of the following are typical water pollutants?

   a) Carbon dioxide gas ❐
   b) Waste oil ❐
   c) Volatile organic compounds ❐
   d) Raw sewage ❐

8. Which of the following are the categories of drains that may be on a commercial premises?

   a) Surface water ❒
   b) Clean water ❒
   c) Foul water ❒
   d) Muddy water ❒

9. Which of the following could be environmental nuisances?

   a) Noise emissions ❒
   b) Light emissions ❒
   c) Carbon dioxide emissions ❒
   d) Oil spillages ❒

10. Which of the following are waste streams targeted under producer responsibility?

    a) Glass ❒
    b) Paper ❒
    c) Lead ❒
    d) Used oils ❒

11. Which of the following are fossil fuels?

    a) Oil ❒
    b) Wind power ❒
    c) Hydro-electric power ❒
    d) Natural gas ❒

12. Which of the following are renewable fuels?

    a) Coal ❐
    b) Solar ❐
    c) Natural gas ❐
    d) Hydro-electric ❐

13. What is an environmental aspect?

    a) The probability of causing harm ❐
    b) Products or services that can interact with the environment ❐
    c) Any change to the environment, whether adverse or beneficial ❐
    d) The consequences of pollution ❐

14. What is an environmental impact?

    a) The likelihood of change ❐
    b) Any change to the environment, whether adverse or beneficial ❐
    c) Products or services that can interact with the environment ❐
    d) The probability of causing harm ❐

15. What is the internationally accepted environmental management standard?

    a) ISO 18001 ❐
    b) ISO 14001 ❐
    c) ISO 9001 ❐
    d) ISO 15001 ❐

16. What does LCA stand for?

    a) Life-cycle analysis ❒
    b) Life-cycle assessment ❒
    c) Long cyclical assessment ❒
    d) Long cycle analysis ❒

17. Does the ISO environmental standard require continuous improvement?

    a) Yes ❒
    b) No ❒

18. Which of the following actions reduce the levels of pollution associated with vehicles?

    a) Using electric vehicles ❒
    b) Increasing vehicle speeds ❒
    c) Reducing the vehicle usage ❒
    d) Increasing the size of the vehicle tyres ❒

19. What is the missing word in the following sequence: Source, xxxxxxx, Receptor?

    a) Vehicle ❒
    b) Pathway ❒
    c) Airflow ❒
    d) Rain ❒

Well done, you have now finished the test. Please check to ensure that you have answered all questions (remember some questions require more than one answer).

Common Sense Guide to Environmental Management

An essential and short guide for those who need to know more about environmental management in the workplace without wanting to spend hours reading dozens of different documents. Whether it's for use alongside a training course or simply to brush up on your knowledge, it's perfect for equipping you with the principles of environmental management.

Friendly and accessible, this Common Sense Guide covers all the main aspects of environmental management in manageable chapters to provide you with the knowledge and understanding you need to look after the environment and those around you.

- Suitable for those with little understanding of environmental management
- Includes questions at the end of each module to consolidate your environmental management knowledge
- Certificate offered to those who complete the exam at the end of the book and return to be marked externally.

Subash Ludhra is a past president of The Chartered Institution of Occupational Safety and Health (IOSH) and considered to be an expert in the field of Risk Management. Having qualified as an Occupational Hygienist, Subash Ludhra now manages Anntara Management Ltd; an international risk management and loss control consultancy business that operates in the UK and overseas.

COMMON SENSE GUIDES TO HEALTH AND SAFETY

Common Sense Guide to Health and Safety at Work
978-0-415-83544-2 February 2014

Common Sense Guide to Fire Safety and Management
978-0-415-83542-8 February 2014

Common Sense Guide to Environmental Management
978-0-415-83541-1 February 2014

Common Sense Guide to International Health and Safety
978-0-415-83540-4 September 2014

Common Sense Guide to Health and Safety in Construction
978-0-415-83545-9 October 2014

Common Sense Guide to Health and Safety for the Medical Professional
978-0-415-83546-6 November 2014